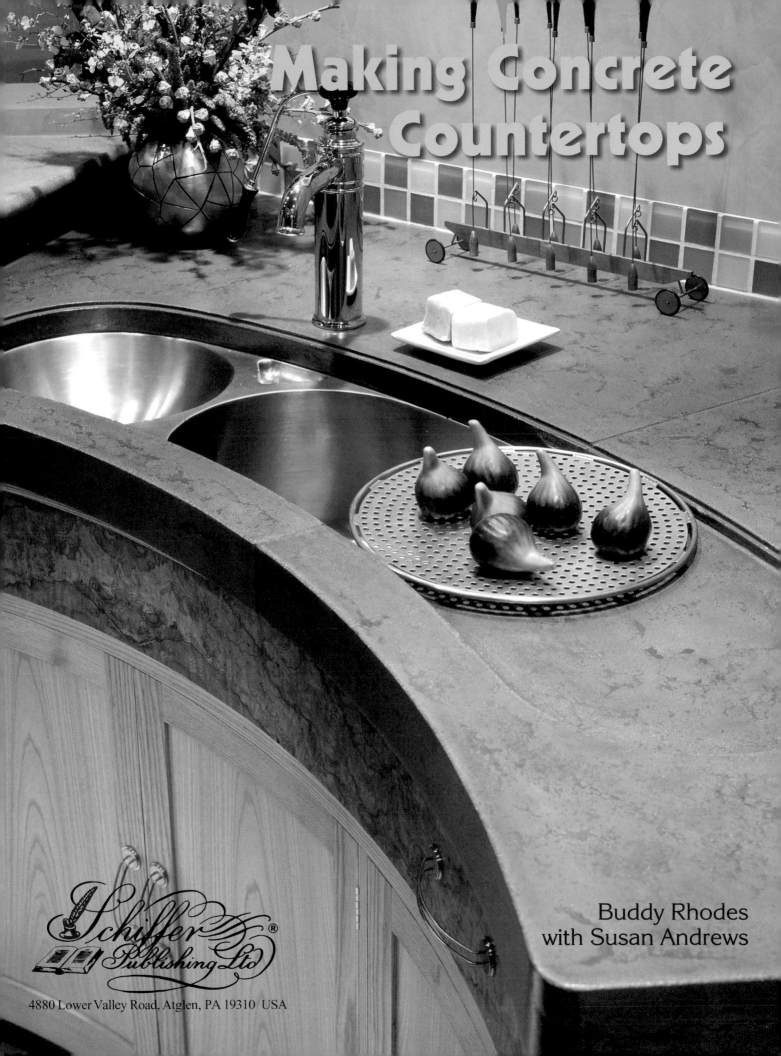

Making Concrete Countertops

Buddy Rhodes
with Susan Andrews

Schiffer Publishing Ltd

4880 Lower Valley Road, Atglen, PA 19310 USA

This book is respectfully dedicated to George E. Orr (1857-1918), "The Mad Potter of Biloxi"

Library of Congress Cataloging-in-Publication Data

Rhodes, Buddy.
  Making concrete countertops / Buddy Rhodes, with Susan Andrews.
     p. cm.
  ISBN 0-7643-2477-2 (hardcover)
  1. Concrete countertops--Design and construction.
2. Concrete construction--Formwork. 3. Countertops--Materials. I. Andrews, Susan. II. Title.

TH6010.R48 2007
643'.3--dc22

                              2006028975

Type set in Geometric 231 Hv BT/Souvenir Lt BT
ISBN: 0-7643-2477-2
Printed in China

Published by Schiffer Publishing Ltd.
4880 Lower Valley Road
Atglen, PA 19310
Phone: (610) 593-1777; Fax: (610) 593-2002
E-mail: Info@schifferbooks.com

For the largest selection of fine reference books on this and related subjects, please visit our web site at www.schifferbooks.com
We are always looking for people to write books on new and related subjects. If you have an idea for a book please contact us at the above address.

This book may be purchased from the publisher.
Include $3.95 for shipping.
Please try your bookstore first.
You may write for a free catalog.

In Europe, Schiffer books are distributed by
Bushwood Books
6 Marksbury Ave.
Kew Gardens
Surrey TW9 4JF England
Phone: 44 (0) 20 8392-8585; Fax: 44 (0) 20 8392-9876
E-mail: info@bushwoodbooks.co.uk
Website: www.bushwoodbooks.co.uk
Free postage in the U.K., Europe; air mail at cost.

Inside This Book

Acknowledgments

This book is a collaboration with my writer-in-residence, best friend, and wife, Susan Andrews. She puts my ideas into words. It's easier for me to put them into cement.

This collaboration would never have been completed without the efforts of our wonderfully kind, prolific, and patient editor, Tina Skinner, who believed in this project from the beginning and coached us cheerfully through the entire process. Without the skilled photography of Douglas Congdon-Martin and Andrew McKinney, our words wouldn't help much. The objective has been to clearly plot out the mini-steps involved in fabricating a concrete countertop with our methods. Doug and Andrew understood this and artfully made it happen.

I am constantly encouraged and held up by my partners and staff at Buddy Rhodes Concrete Products, each of whom performs the job of five people. June Mejias mothers all of us, customers and employees alike, at home base; Matthew Mondini expertly juggles technical support calls and visits, warehousing and plaster molding, while Josue Cruz assists, packs, and finishes concrete with a vigor reserved for the young and intense. Jim Mason helped us move from fabrication to manufacturing and distribution, finding homes for our products across the country; Rich Rhoades and Matthew Newman go out on the road to spread the word on how we use them. Lorna Villa keeps us on track while Steve Schatz, Mark Gunther, and Steve Massocca keep our doors open with their vital knowledge and multi-leveled support.

Heriberto Esquivel, who grew up at Buddy Rhodes Studio, is now a teacher, contractor, and masterful concrete artist on his own. The owner of Concrete Concepts and Design, as well as head instructor for BRCP San Francisco workshops, "Beto" generously coaches his future competitors who may never catch up to his extraordinary skill and artistry – a nearly impossible feat. As our largest single user of BR concrete products, he provides a busy laboratory for research and development.

I am ever thankful to those in my family and community who have inspired, supported, cared for, fed, taught, befriended, and advised me during my research and development years. Without these people, I would not have been able to continue playing with my mud and sand for long enough to make my pies. I must therefore gratefully acknowledge the following people and places: My Mom & Dad for always supporting me; Val Cushing and Daniel Rhodes of Alfred University, The Rhode Island School of Design, and Richard Shaw at The San Francisco Art Institute for teaching me; David Sheff, Carolyn Zecca Ferris, Peggy Knickerbocker, Pamela Barish, Joel Torrnabene, William Passarelli, Liane Collins, and Bobby and Sheila Fetzer for believing in me; Laddie Dill and Stephen Andrews for inspiring me; Ron Mann, Dan Friedlander, Topher Delaney, and David Baker for finding me; David Hertz, Fu Tung Cheng, Mark Rugero, Mike Miller, Tom Ralston, Steve Rosenblatt, George Bishop, and Jeff Girard for pioneering with me.

Lastly I am indebted to all of the architects, designers, and trend setting homeowners who paved the way, and the many gifted concrete artists all over the country who now carry on the work.

Photography

The very patient and gifted photographer Doug Congdon-Martin took all of the fabrication shots in this book; the on-site templating and installation images were captured by the expert San Francisco digital photographer Andrew McKinney, as were the final images of the Scott Street apartment which served as our laboratory. Most of the images of kitchen and bath vanity tops were supplied by Andrew or Doug, along with David Duncan Livingston, and Ken Gutmaker who has taken many of our most prized photographs of Buddy Rhodes Studio projects.

Designers

I have had the privilege of working in partnership with some very talented designers who, with their generous clients, have chosen my surfaces for their kitchen and bath projects. The names of the esteemed designers and photographers whose work is represented in these pages can be found in the Credits section.

Introduction
Turning Powder to Stone

I figure when you wake up in the morning you need to find something to do that you believe in. Feeling directionless after high school on Long Island, I discovered clay. Where had I been? It was heaven and I could make magic there. I could turn powder to stone. I could make mud pies and transform them into vessels.

It took me a year to learn to center. The wheel was captivating. I started making bowls, plates, and vases – functional things. My vessels turned into plates. Plates evoke feeding the family. Finally, I too was centered.

At Alfred University in upstate New York they took pottery very seriously. The first assignment was to develop my own clay body or formula. I felt like I was learning to make mud and grow life. Always fascinated with geology, I was combining science with art, adding water to mineral powders and making things grow out of it.

It was the sixties turning into the seventies and the simple life of living on the land was celebrated: Back to basics – Grow your own food – Weave your own fabric. I became a potter and moved to an old farmhouse on Lake Cayuga, touring crafts fairs in the Northeast. But the simple life as a potter was cold, and not even the kilns I built and the friends I took in could keep me warm in my beautiful old house. I joined several of my favorite ceramic artists in California and enrolled in a graduate program at the San Francisco Art Institute in 1979.

With the physical move came a shift in my forms. I turned from vessels and plates to slabs with which to build. My form became bricks fired all night in the basement of the Institute. I developed a new scale of work. Bricks became tunnel structures, which became clay houses in the school's courtyard. I lit them on fire and blew them up; it was art school after all. I kept needing more bricks and there weren't enough hours in the day and night to do all the firing.

I needed a self-firing, self-hardening mud. I always enjoyed taking powder and turning it into stone. Cement is a powder made from pre-fired and crushed limestone paste. When you add sand and rocks, you get concrete. Experimenting with mixes, I learned to mold my bricks with concrete, just as I had molded all along with my clay body. Concrete, like clay, took on a life of its own. No more firing. It, too, was magic.

I liked the veins I created when I pressed the material against the mold. I liked making multiples. Out of art school, I began making pieces of furniture, slabs, planters, and rounds. Fabricating furniture, like the pottery pieces a decade before, was making objects for people to use. I still wanted to put craft into people's homes.

My first concrete counter was in my own home in the mid-eighties, cast in place and burnished by hand for hours, returning throughout the night as the concrete set. This way I could put my

yet they have the hand of the craftsman about them. I like the veined surface I make because it looks like it has always been there.

Visiting Europe for the first time on my honeymoon in the early nineties, I fell in love with the stonework and the age of the surfaces. People had been living with nature for a long time. Buildings had huge cracks. Archways in Gubbio were held up by tree trunks! Steps were worn by centuries of foot traffic. There were fractures, patina, cracks, and scratches showing the ways generations had lived and used surfaces. Centuries of grandmothers had all chopped their onions right there.

pottery to work every day and it wouldn't be sitting on the shelf like my vessels or in a gallery like my furniture. A few daring architects and designers started hiring me to build concrete counters for them on site.

Soon I started pre-casting them, much more satisfying for me because it harkened back to the fabrication of multiples in my pottery days; it gave me more control. It allowed me to experiment in my shop, to shape molds and pack them and color the concrete and make a mess in privacy.

I still get a lot of satisfaction from seeing my countertops over time. I like the way the trowel leaves its signature or path on the surface, the leather-like effect of a sealed finish. I like the way the edges are straight, the designs are tight,

It's really great to hear people exclaim, after years of living with my concrete counters, that they've loved them. And when they buy a new house they want concrete counters again. But it's not just the material. It's the craftsmanship. Workmanship makes the difference. It's the feeling, the process of tuning and adjusting. It's not necessarily easy, but it is very rewarding.

The purpose of this book is to help you turn powder into stone and make magic that people can use every day with love. It is an opportunity I'd like to share with you.

Demonstration Projects

The projects in this book are not representative of our grander projects. However, in one small city apartment space we were able to illustrate three compact examples of our most commonly requested surfaces: pressed, hard-trowelled, and ground. Heriberto Esquivel, my former BR Studio foreman who now heads his own company, Concrete Concepts and Design, contracted this San Francisco project, which the owner planned to put on the market after its renovation. It required a pressed, or veined, finish kitchen counter and island in a warm gray tone, a trowelled double-sink master bath vanity top, and a guest bath vanity with a ground-polished surface. In the following pages, you will be able to follow the project from start to finish, with intimate, step-by-step imagery. Some steps in the process are common to all three surfaces, e.g. templating, mold making, and mixing. Where the process steps diverge according to countertop types, you will be

able to follow those methods in separate sections. Each process along the way from start to finish is designed to show mini-steps that you can follow sequentially, or skip to as desired. So bear with me if you already know about all the tools, or how to make a template (very similar to how its done for other stone and solid surface applications) and refer to your chosen chapter or walk through the process with me step by step.

Getting Ready

Tools

You will need the following tools for pre-casting concrete countertops as described in this book:

- A cement mixing device: hoe and wheelbarrow, hand-held mixer, mortar mixer, or cement mixer
- One or more strong level tables, preferably with six legs
- Rubber gloves, safety glasses, dust mask
- Glue gun with glue sticks
- Marking Pen with two colors
- Table saw or skill saw
- Jig saw
- Chop saw
- A cordless screw gun
- Drill press
- Counter sink bit
- Level
- Square
- Wire cutters
- Assorted trowels (a wood float and different size stainless steel trowels)
- Putty knife
- Sander
- Sanding wheel
- Tape measure
- Radius template
- Vibrator
- Band saw

Materials

Here is your materials list:

- 4 x 8 foot sheets of melamine, depending on the size of the project or 1/8-inch glu-lam plywood

The projects in this book are primarily made with Buddy Rhodes Concrete Products, including:

- Concrete Mix
- Color to mix with water
- Cement Paste to fill voids or air holes as needed depending on method
- Penetrating Sealer
- Satin Sealer
- Beeswax
- Edge forms – handmade or purchased

These commonly available materials are not distributed by Buddy Rhodes Concrete Products.
- Fiber, to add tensile strength to sinks and unusual shapes
- Reinforcement wire, expanded galvanized wire mesh or ladder wire (Ladur™ by Dur-O-Wal™)
- Drywall nails
- Assorted drywall screws
- Sheet of 1 1/2-inch thick Styrofoam
- Heavy plastic tape
- Spray adhesive
- Bondo™
- Rubbing Stone (carborundum)
- Clean rags

Things to Know Before You Start

- The processes detailed in this book will take 7-10 days from start to finish.
- Prepare to start the project the day before by setting up tools and materials.
- Set up forms the day before you intend to mix the concrete.
- Double check the molds against mea-surements or template before adding concrete.
- Always mix and pack forms in the morning. This way you have time to work the concrete as it sets without staying up all night.

Making Samples: It is a good idea to cast a sample project before you get started on a countertop you plan to install. Even if you are familiar with casting in place, or working in concrete with other applications, it is always helpful to get your hands in the mix and make a practice piece first. In addition to the value of practicing the techniques in advance of making your perfect countertop, it is also important to make samples of the color and finish you intend to produce. This preview will demonstrate to you and your client what can be expected when you finish your expanded project.

A Pressed Surface Project

A pressed surface mold is made upside down. The cement is pressed into the upside down mold against a smooth melamine surface. This process creates the highly desirable veining – voids that have been backfilled. The effect is one of age, and organic origin, like timeless stone.

With more traditional concrete methods we hear of "pouring" concrete. Pressing concrete is different in that it's dryer, more malleable. It works with three-dimensional projects too. Without the heaviness of pouring into a double-sided mold, I pack my dry mix up vertical walls of a planter mold the same way I construct a countertop.

You have to think differently when learning the pressed method. Imagine constructing a clay coil pot by hand, as opposed to pouring plaster into a mold. There is more crafting to it. That's the direction I came from when I first started working with concrete. Moving on from cups and bowls, I began packing clay into plaster molds and making bricks. To avoid firing, I switched to using concrete the same way. The bricks evolved into concrete furniture, once again working with function in the home. The look came from how I worked with clay. The road from bricks to countertops was bringing me home again. I believe that is the appeal of concrete countertops: warm and earthy, they remind the user of handcrafted ceramics.

Unlike when pouring concrete, with the pressed method, bug-holes are welcome. You can subtly infill voids with a tone similar to the base concrete matrix, or add contrasting colors, singly or overlaid in up to three or four combinations. The variation and the possible options offer a limitless palate to the concrete fabricator.

Making the Template

Templating is the simplest way of measuring, and the first step to building a custom mold for any pre-cast countertop project. It is a low-tech process of physically laying down strips of plywood in place exactly where the concrete slab is to go, carefully gluing them with the help of a glue gun and finish nails, and marking them with important details. Cardboard or paper templates do not work as well because they are neither firm enough nor flat enough to assure an exact pattern with a crisp clean edge. Drawings never represent reality as well as a template made on site fitted to the wall. After all, the wall may be slanted or curved.

While on site, the template maker can determine the best place for seams, make sure there is room for the faucet holes, and identify the center of the sink.

It is better to have more information than not enough. You will be taking this pattern back to the shop. There it becomes your connection to the future home of your slab. Whether you will be casting upside down or right-side up, label the top of the template with a felt tipped pen, as well as the finished edges and any other important details. I even like to have a digital photo of the template on site to remind me of the orientation of the countertop when I'm making the mold back at the shop. Memory fails, and it's particularly helpful when a different person makes the template than the one making the mold.

You may choose to make your counter a little bigger when you aren't completely sure of your measurements. You can always cut a slab down, but you can't stretch it to fit! Check at every step of templating and mold making. At our shop we say, "There's a mistake here somewhere and it's our job to find it." The best time to find a measuring error is before you cast it.

I have made all the mistakes, some of them several times!

Aside: Correct measuring and templating are the result of basic skills plus repeated review. Don't check your measurements twice; check them three times. Similarly, repeatedly make sure that the template is placed with the appropriate side up in the mold, and that all the necessary notes are correct and understood.

The first step in creating a concrete countertop is to build an accurate template made of strips of 1/8-inch luan. The strips are cut from 4 x 8 foot sheets, 2 1/2 inches wide. Roughly cut the strips to fit the rim of the countertop.

Measure the overhang from the cabinet frame. In this case it is 1-inch.

A utility knife and a square are used to cut the pieces of luan square.

Tack the strip in place using drywall nails.

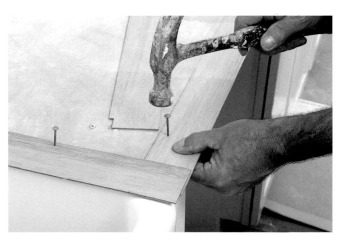

On the front edge piece, place the template to create an overhang that is flush with the cabinet's door front. A square helps in alignment.

Align the front of the strip to the correct 1-inch overhang and tack in place. Next, lay in a strip flush with the back wall and tack it in place.

Here is a common problem calling for a custom fit. The side of the cabinet beside the countertop has an edge that protrudes into the countertop. Lay a piece of template strip in place and mark the shape to be cut.

Note: if this were not a "country-style" sink with the deep front apron integral to the design, we would make a template for the front edge of the sink as well. In this case the template will be in three pieces, each side and the back, with the seams along the side edges of the sink. In the other it would be a four-piece template, the two sides, the back, and the front.

Trim it out with the knife.

Test the fit and make any necessary adjustments.
At the sink you need to measure for an overhang. Start with the right side of the sink. Measure at the front corner.

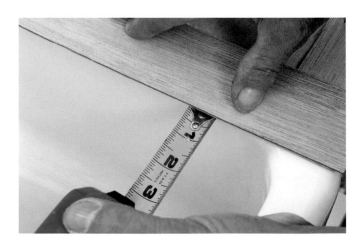

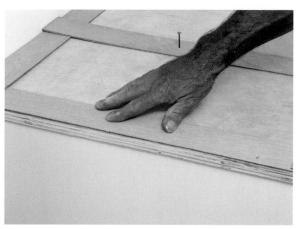

 And measure at the back corner. When the right side template is aligned, tack it down

At the stove opening, the countertop is flush. Align and tack in place.

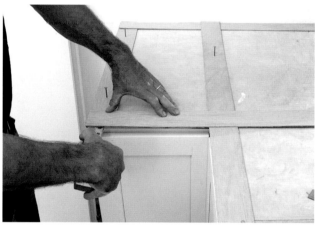

 At the left of the sink, measure an overhang of 1-inch for the countertop.

Running a strip the entire length of the countertop creates a more structurally sound template.

 When the overhang is correctly aligned, tack it in place.

Check with a square for a 90-degree angle and make adjustments.

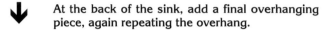

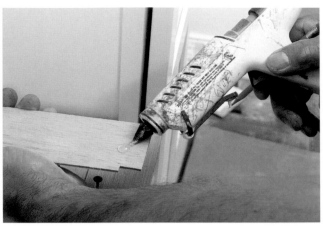

At the back of the sink, add a final overhanging piece, again repeating the overhang.

Apply glue from a glue gun.

Alternatively, if you do not want an overhang you can align the template around the sink so it is flush with the inside walls of the sink. Align at one point.

Glue the end pieces to the front and back edges.

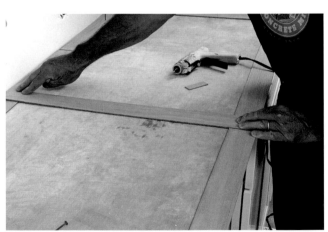

The template is ready for gluing.

Add crosspieces as needed to add support to the template.

Where pieces are tacked down to hold position, as here by the sink, just lift up the edge and apply the glue.

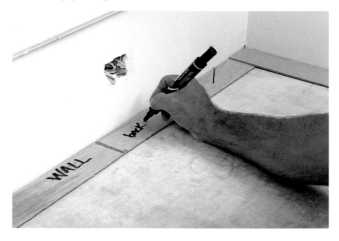

When the gluing is complete, mark as much information on the template as necessary.

Nearly complete: I use "FE" to notate "finished edge." Small diagrams indicate positions and overhangs.

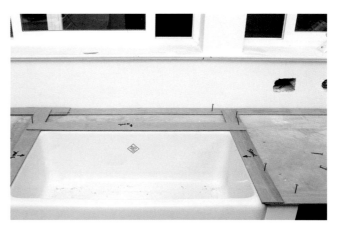

Aside: Sinks and Faucets

Sometimes the designer of the counter space doesn't take into account the space needed for the sink and faucets. It is important to make sure that the sink will fit into the cabinet as shown in the drawing. You might need to move it forward or backward. I like to bring the sink back to the shop if possible. New sinks in the box often have a paper template with measurements of the sink, which is very helpful to have when making the mold.

At the back corners of the sink, glue a diagonal piece that overhangs the bowl.

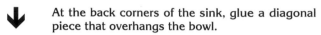

Holding a pencil flush to the side of the bowl, mark the radius of the bowl at the corner. This is important to create rounded corners during the molding process.

We need to cut the template in half to bring it home. Mark the center and add support pieces on each side of it.

Cut a temporary seam.

The backsplash above the sink needs openings for outlets. Again, it is best to build a template so the holes can be part of the tile design. Cutting the holes in the concrete is tough work. Tack the upper and lower crosspieces in place.

Glue strips on each side of the outlet opening, aligned with the electrical box.

Glue a crosspiece aligned with the bottom edge of the box.

Because the boxes are standard dimensions, defining three edges gives enough information to build the mold.

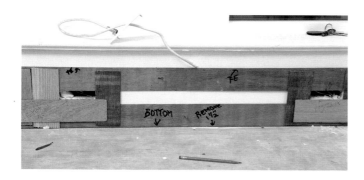

Carefully mark the template. Include a reminder to remove the thickness of the countertop slab from the backsplash template – in this case 1-1/2 inches that will later be cut from the bottom of the backsplash template on a table saw.

The island is two-tiered and requires two templates.

The lower template needs to work around two posts, one at the back...

and one at the end.

The upper level has only the end post to work around.

Making the Mold

The mold is made with 3/4-inch melamine-covered particleboard. The finished edge of the kitchen counters will be 2-3/8 inches thick, while the main body of the counter will be 1-1/2 inches thick.

Cut pieces to size on the chop saw.

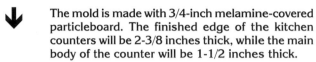

4 x 8 foot sheets of melamine particleboard are ripped for ease of handling, first in half to make it a 2 x 8 sheet.

The template was cut to make it easier to remove from the home. It will be cut in three pieces, and in order to determine the layout, the pieces need to be put on the melamine table side-by-side. They are reversed, all instructions transferred to the opposite side, then aligned carefully and tacked in place.

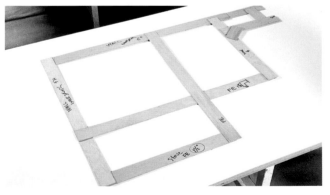

These strips are then ripped into quarters ...

The kitchen counter is done in an upside down mold. This is the template for the left side of the counter, between the sink and the stove, as developed onsite.

 We turn this template upside down and mark it "Up in Mold" or U.I.M.

With this done, we transfer all the notes made onsite to this side of the template.

Tack the mold in place on the melamine tabletop. Because this countertop is poured upside down, it is very important to work on a new, flawless piece of melamine. Any flaws in the table surface will show up in the countertop.

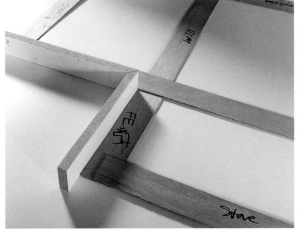

Begin to build the mold wall. For this counter the exposed depth at the front edge will be 2-3/8 inches, so the melamine mold strips for the exposed edge need to be that deep. Internal right angles in the mold walls, such as this one, should be mitered. This avoids the exposure of raw particleboard, which will leave an unattractive finish. Other joints can be butt-fitted.

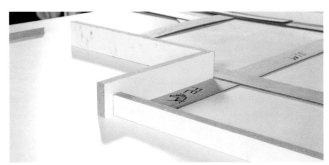

 The wall and stove edges will be just the thickness of the top, or 1-1/2 inches. The mold walls for this, then, need to be 1-1/2 inches.

Cut pieces to length and dry fit them. Mark where the screws will go down to attach the edges to the melamine and pre-drill the melamine on a drill press. Create countersink holes with a countersink bit on your drill.

There will be seams in the back of the sink, dividing it into three segments, so the mold wall does not need to extend to the end of the template.

Glue it place behind the apron wall at each end.

Attach the mold walls to the melamine. It is important to choose the correct length screw. They should be long enough to go into the melamine, but should not penetrate through the other side.

The underside of the apron is formed by these 1-1/2-inch pieces, mitered at the joint. Mark holes for screws so they don't align with previous screws. Drill the holes and countersink them.

To create a 3/4 inch thick return edge, we use a 3/4 inch block at the junction between the 2-3/8-inch front and the 1 1/2 inch front and the 1 1/2 inch body.

You will fit and cut a lip to determine the thickness of the finished edge wall. Pre-drill and countersink screws in the lip, being careful that they do not hit previous attachment screws in the wall.

 The counter section to the right of the sink is done in the same way. After marking U.I.M. on the opposite side, transfer all the marks from the template.

 Again, apply a 3/4-inch block at the junction of the walls of different heights.

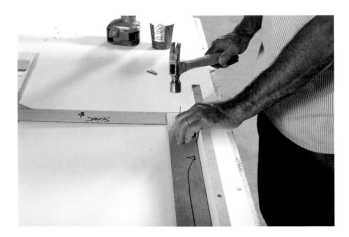

 Tack the template to the melamine.

The underside of the finished front wall is formed by this 1-1/2-inch piece.

 Dry fit the mold walls, and then screw them in place. Again we will have a 2-3/8-inch front edge with the other edges being 1-1/2-inch, the thickness of the countertop.

Again, attach a lip on the finished-edge wall. Mark holes for screws in the top board so they don't align with the screws underneath.

↗ Drill the holes and countersink them.

Attach the strip to the edge of the mold.

Making the Sink Knock-Out

⬇ At the jobsite we drew the template so the edge of the sink cutout was flush with the sink walls. The owner later decided to have the countertop set back from the sink edge by 1/2-inch, which we notate and adjust on the template.

Transfer this to the sink template and cut the corners. ⬇

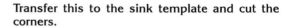

 Take the measurement of the sink. Transfer it to a piece of luan and cut it to size. Then mark the centers of each side and draw centerlines on the template.

A final check is made to ensure the sink template fits the countertop template.

Turning to the corners of the sink marked at the site, we use a radius template to find the best match. In this case it is 3 inches.

Secure the sink template to 1-1/2-inch Styro-foam™ with a drywall screw.

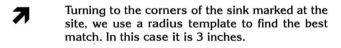

 Reduce the Styrofoam to the correct size on a sanding wheel.

Make small cuts in the corners of the tape.

 Carry the centerlines over the edge of the template and onto the Styrofoam sides. Remove the luan and draw the centerlines across the Styrofoam.

Fold over and seal the edges.

 We use tape to finish the edges of the Styrofoam for smooth casting. Working on a flat table, start at the center of one side and apply tape to the edge. Continue around the sink template until the tape meets. Cut it to make a flat joint with no overlap.

We want the seams between sections of countertop and the piece behind the sink to fall where the counter is straight, 1/2-inch in from the end of the curved corner. Since the radius of the corner is 1-1/2 inches, I add another 1/2-inch to determine where the seam will fall, and mark it on the knockout.

A piece of clear tape over the top and bottom surfaces will keep the trowel from catching in the tape.

Cut the Styrofoam sink mold on the table saw, taking 2 inches from each side. Retape the end piece of the Styrofoam sink mold, carrying the tape around to the newly exposed surface.

Rejoin the halves of the countertop template on both sides with hot glue. This involves removing one of the countertop templates from its mold box.

Reseal the tape on the edges.

On the side of the template that is still in place in the mold box, mark the position of the seam, 2 inches in from the edge.

Carry the line across the back of the counter template.

Return the other template to the mold, mark it 2 inches from the side of the sink, as before, and glue a support piece on.

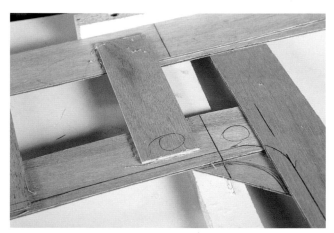

Add a crosspiece to support the end of the middle counter template. It should be near the line you just drew.

Remember to mark opposite sides of the cut to keep the orientation from becoming confused. You might mark one side "X" and the other "O".

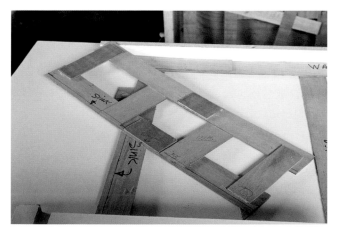

Cut the template at the seam.

We now have three pieces to the kitchen counter. Two large end pieces and a small centerpiece behind the sink.

 Now that the template has been divided into three sections, we can apply the end mold wall to one mold…

 … and the other.

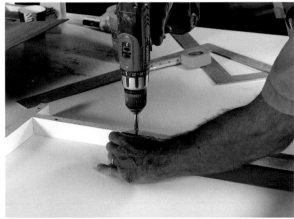 The mold walls are now complete and we can remove the template.

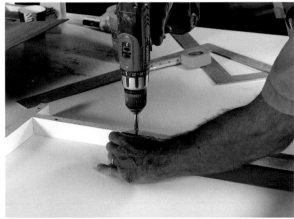

Attach one foam sink edge form in the corner of  the mold.

Do this in both molds.

The Backsplash Mold

Tack down the template for the section behind the sink.

Build the mold wall. This mold is a simple rectangle.

The three pieces of the countertop need to align, so I measure and mark and measure and mark.

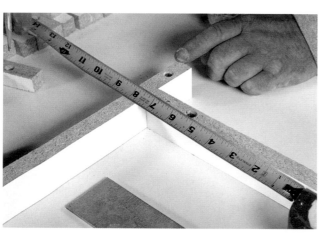

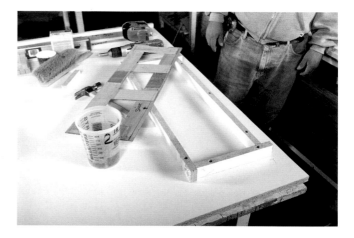

When the measurements all agree, put in the final wall of the center section mold.

Note: A backsplash can compensate for a counter slightly too narrow by covering the gap in the back. However, if it is too wide, it is difficult to fix without recasting.

Clearly mark the mold.

We are going to make the backsplash
in 3/4-inch tile segments.

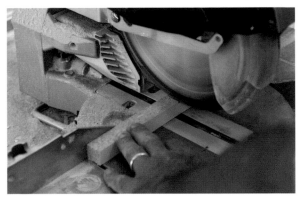

All the mold walls are 3/4-inch strips of 3/4-inch
melamine. Attach a long piece to the melamine
backing board to act as one wall, using a straight
edge for alignment.

The backsplash in the kitchen is 5 inches tall.
Cut mullions 5 inches long from 3/4-inch strips
of 3/4-inch melamine. These mold walls will
separate the tiles.

Divide the length of the backsplash into the desired number of tiles. Remember that a 3/4-inch
allowance needs to be made for two backsplashes
that butt-fit in a corner.

Install a mold wall at that point. Continue with this
process for each backsplash tile.

Attach an end wall at one end of the long backsplash mold using a square for alignment. Measure
for the length of the first tile.

Two tiles are at the right of the sink and have outlets. Transfer their location from the backsplash
template made onsite.

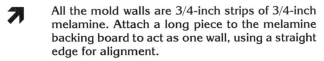

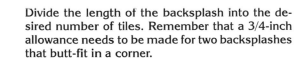

Making Outlet Knockouts

Blocks of 3/4-inch Styrofoam are cut to the size of the outlet boxes.

Wrap the edges with tape.

Fold over the tape.

Align the outlet space holder and screw in place.

Continue with the next tile mold.

Add another long wall at the top of the tile molds, clamping it to a straightedge to assure straightness. As shown in the foreground, the template was cut on a table saw to eliminate 1-1/2 inches for the thickness of the countertop slab.

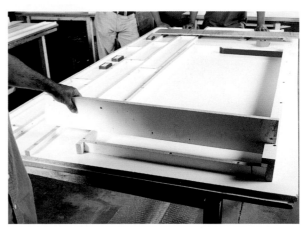

 Screw the mold wall in place.

 A piece of melamine is cut to size, with a notch that fits the profile of the apron overhang. This will ride on the front and back mold walls and form a screed for flattening the underside of the countertop.

 Continue with the tile mold making process. Using the top of one tile mold as the base for another tier.

 Rubber faucet hole knockouts are wrapped in tape.

 Fill the screw holes with modeling clay before casting with concrete. This makes it easier to remove the screws afterward. This is only done on upside down molds, not right side up ones, where the trowel might blend modeling clay into the surface during finishing.

 When you have located the center of the fixture, it helps to make a pilot hole through the foam and just into the melamine. A screw will do. This way you can feel when the placeholder is in the correct position.

Rounded Corners

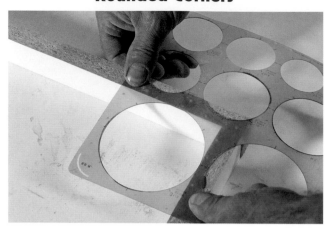

Sometimes the design calls for a rounded corner that must be built into the mold. In our kitchen project, we created rounded corners for the island.

Apply tape so the Bondo™, a two-part putty, doesn't stick to the wood.

Select the radius of the corner. Transfer it to a block of wood of the same height as the mold wall, in this case a 1-1/2-inch chunk of 2 x 4.

Trim off the edge of the tape.

Cut it out on a band saw and sand smooth for this result.

Insert a screw into the corner so it protrudes about 1/2-inch, but not into the block. This will anchor the Bondo mold in place.

Apply beeswax to the block to help in its release.

Move the block to the side and butter the corner of the mold with Bondo.

Add the catalyst to a small amount of Bondo.

Move the block in place and work more Bondo into the gap.

Mix it in, following the manufacturer's instructions.

Tap the mold frame to move the Bondo deeper into the mold.

 Let it cure. This usually takes about an hour, but may vary with conditions.

Butter the corner and smooth.

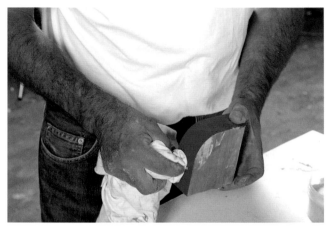

 Removing the block, you can see that there are voids.

Rewax the block.

 A close-up look.

Push the block in place and top it off with more Bondo.

 Squeeze the mold tightly and clamp the block in place. When it has set, remove the block.

Working down to a finer paper, feather the Bondo to the melamine mold wall.

Sand the surface, beginning with coarser paper.

A skim coat of Bondo fills in the voids and smoothes the surface.

 Remember to sand away excess Bondo from the mold wall.

Sand the surface again, using a finer grit.

Preparing Reinforcement

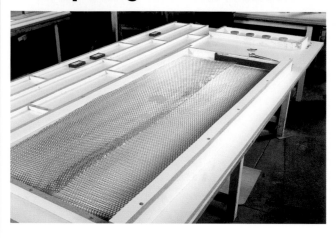

A second piece goes the length along the back.

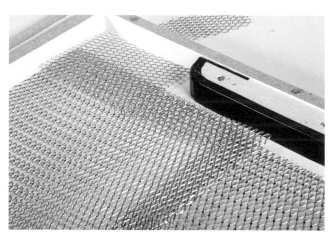

A third piece fills the little space behind the sink mold.

Shellac the surface to fill any minute voids. If there are any rough spots they can be sanded out.

The corner should be coated with water based form release before filling.

Single pieces are cut for the tiles.

 For the main countertop, wire should be cut to come to about 1-inch from the edge and narrow enough so two pieces overlap a couple of inches in the center. Working on the kitchen counter, this piece is bent to fit into the apron.

Remove the wire mesh, being sure to keep it organized. It will be placed in the concrete after the first layer is laid. Clean the mold using compressed air or a vacuum cleaner.

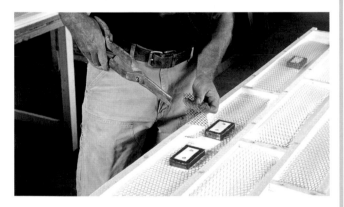

Spray the mold with a water-based release agent. We use Duogard II™.

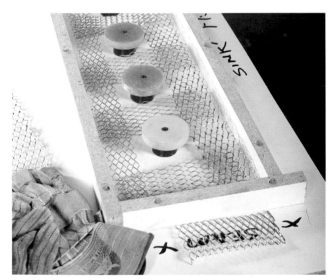

Overlapping pieces are used in odd spaces like around the outlets in the backsplash tiles and/or around the fixtures behind the kitchen sink.

Spread the release agent all around the mold, covering every surface using a clean cotton rag. Now you're ready to mix in the morning!

Mixing the Concrete

There is a progression of low tech to higher tech ways of mixing concrete for concrete countertops. I started with a six-cubic-foot tray and a hoe. I have used a hoe and wheelbarrow too. At my first studio, I had one employee helping me. We would stand at either side of a mixing tray and pull or push our hoe through the concrete mix, back and forth, until it was the right consistency. I still do that sometimes in my eleven-cubic-foot tray. More commonly now, when I'm just mixing a couple of bags, I use a hand-held Perles ME 140 cement mixer with a paddle attachment, commonly used for mortars, grouts, and cement.

For bigger jobs, I use a gas powered mortar mixer with a paddle attachment. I graduated from my heave-ho tray technique to a drum mixer. Drum mixers will work, but keep in mind that they take more time, and are designed to rely on larger aggregate and gravity to do the mixing. I do not have large aggregate in Buddy Rhodes Concrete Product Mix™ (BRCP), so the mortar mixer is preferable.

The most important thing I tell my students is to go for consistency when mixing, rather than matching a written recipe. Water content requirements may vary by geographic area due to air moisture and regional variations in blending, even with a single recipe. White sand, for example, may differ slightly in mineral content from east to west.

The most common error made in casting concrete countertops is the addition of too much water. The goal is to use as little water as possible. Those familiar with pouring concrete for more common flatwork such as patios, driveways, and floors often make the mistake of assuming that they know what concrete should feel like, that it should "pour." I prefer to say we "pack" our molds, rather than "pour." Think wet enough to hold together – just. Think of cookie dough for concrete pressing in molds upside down, and a little looser, like oatmeal, for hard-trowel and polished finishes. I recommend holding back a few cups of water when mixing initially, and adding the final moisture if you need it. Less is more.

I like to start this process first thing in the morning, so I have all day to work with the concrete as it cures. It helps me to avoid the late nights babysitting setting cement.

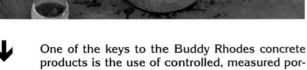

One of the keys to the Buddy Rhodes concrete products is the use of controlled, measured portions of concrete and pigment. The concrete is tinted using liquid color, thoroughly mixed with water, following the instructions on the package.

Add the mixed colorant.

Be sure to use all of the liquid color. It helps to rinse the container with water.

Mix thoroughly. Monitor the appearance of the cement. You'll want a dry, cookie dough-like consistency for pressed countertops, while a trowelled counter material should be more like oatmeal.

In a cement mixer, add the concrete.

Ready for the molding process, concrete is poured into a wheelbarrow for transport.

If you are doing a hard trowel or ground finish project, skip ahead to page 57.

The Press

For the pressed surface countertop we work upside down with the top against the worktable. Essentially it involves pressing small clumps of concrete into the mold one at a time. The concrete is mixed to the consistency of cookie dough or clay.

The basic technique is to grab a clump of concrete and break off a piece with your fingers. This piece is pushed into a mold with the fingers of the other hand, using enough pressure to spread it out, but not enough to smooth it completely. The goal is to leave some voids in the surface of the countertop, that is the surface that is against the worktable. The process is cumulative, one piece at a time. Gradually the pieces will cover the surface, about one half the thickness of the finished slab. Use your fingers to press one clump of concrete into another, creating a bond.

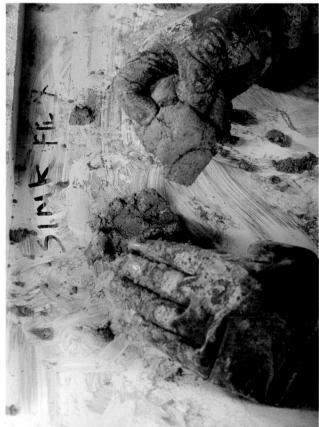

It might be a good idea to start on the tile molds for the backsplash. I begin at one end and work toward the other.

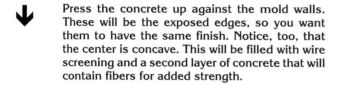

Press the concrete up against the mold walls. These will be the exposed edges, so you want them to have the same finish. Notice, too, that the center is concave. This will be filled with wire screening and a second layer of concrete that will contain fibers for added strength.

Because we want to have crisp edges, we apply the concrete generously into the mold and then go back to press it into the edges with considerably more pressure than we used on the flat surfaces.

The same basic technique is followed on the countertop itself. Start at one end and fill. It is important to overlap the clumps of concrete and turn them into place. If you line them up too regularly you will have unnatural lines in the countertop when it is turned over.

Go back and build up the apron area.

 Build up the edges and around fixture holes. polypropylene fiber is added to the mix of the concrete's second layer for strength. It is first soaked in water and then mixed into the concrete along with a little bit more water to create a slurry. Remember, we are now working on the underside of the countertop, which will not be exposed, so the fibers will have no aesthetic effect.

See page 81 to learn more about reinforcement.

 With a layer of fibrous concrete in place, set the mesh and gently trowel more concrete on top. Begin with the front piece, bent for the apron. Don't apply too much pressure or you will undo the surface effect.

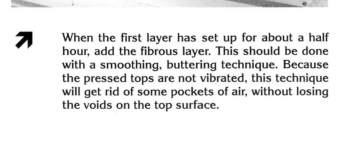

 Here you can see how the mesh has been bent up to fit in the overhang or apron.

 When the first layer has set up for about a half hour, add the fibrous layer. This should be done with a smoothing, buttering technique. Because the pressed tops are not vibrated, this technique will get rid of some pockets of air, without losing the voids on the top surface.

Cover the wire mesh with more concrete.

Work the surface with a putty knife. At this point you are setting the mesh and applying enough pressure to give a solid top, but not so much as to remove the gaps and space that will make the top surface interesting.

Screed the surface using the piece of melamine board cut to fit the mold profile. A back-and-forth motion works well, letting the screed run on the mold walls.

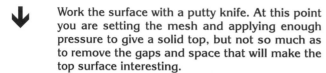

With the mesh embedded, top off the mold with concrete.

Fill in any voids as you go.

Pay attention to the backside of the apron, making sure everything is all pushed in and crisp.

Work the concrete in with your hand.

 Then go over it again with the screed.

Flatten down the cement with a wood or fiberglass float.

 The backsplash tiles are done in essentially the same way. After placing the wire mesh, fill the tile with concrete.

On the pressed countertop I go directly to the metal trowel and use long, smooth strokes. This is the underside of the countertop, so what I am after is a surface that will sit flat on the cabinet.

Pay special attention to the corners.

↗ The surface does need to be flat and smooth.

The same is true of the backside of the apron. Work the backsplash tiles in the same way.

Curing & Finishing

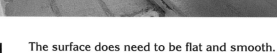

⬇ Covering the countertops overnight slows the evaporation process and provides more even drying.

The forms may be dismantled simply by removing the screws. ⬇

↗ After it cures for twelve hours, the plastic is lifted. Water from condensation is trapped on top, but the countertop is set and ready for de-molding.

Remember to remove screws from the plumbing placeholders. ↗

 The forms should just fall away

 A tap or two on the table with a rubber mallet will help release the tile molds.

On the apron, it is best to clean up the casting before removing the form. We use a carborundum stone. The edge of the form is a good guide for the stone; it takes only a light touch.

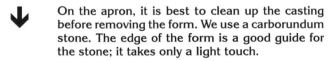

 The sink mold is still a little snug, but instead of prying it out in this position, and risk breakage, I will wait until the top is turned over so I can push it down more safely.

Continue with the form removal. On the apron, the top piece is removed, then the wall.

Turn the countertop over.

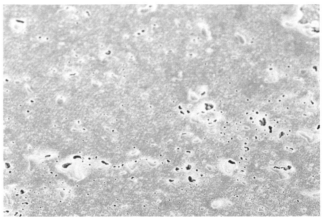

 Because the apron edge hangs down, it is safest to rest it on 2 x 4 blocks for better air circulation.

This is the pressed surface as it comes out of the mold. You can see some lighter borders around the voids. This happens when the wetter concrete used in the application of the wire mesh seeps down and fills the voids from the initial pressing. We will take care of these in a few minutes.

Lower it gently, pivoting on the straight edge, not the apron, which is weaker.

First remove any Styrofoam mold pieces.

 Allow the apron to overhang the table edge just a bit for easier access.

Lightly scrape with a putty knife to remove any remains.

Filling the Voids

 A light sandblasting will remove the "halos" around the voids.

 Blow away any dust or debris created by the clean up.

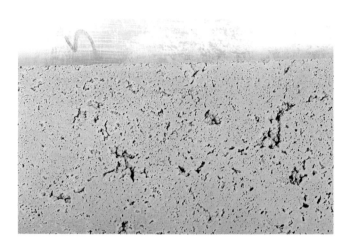

 The result.

 The fill consists of acrylic additive and Buddy Rhodes paste. Mix thoroughly to the consistency of yogurt.

Alternatively you can use a wire brush, but this can discolor the concrete and, on a darker color, can leave noticeable marks.

You want to work on a clean surface, so it is not a bad idea to go over it once again with a brush

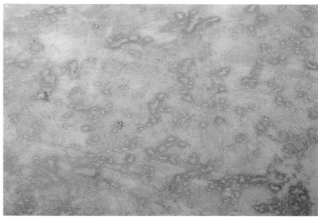

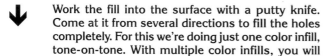

Work the fill into the surface with a putty knife. Come at it from several directions to fill the holes completely. For this we're doing just one color infill, tone-on-tone. With multiple color infills, you will want to leave some voids for the next color.

As the fill dries it shrinks somewhat, leaving depressions and voids.

Work your way down the countertop. You want to make this as smooth as possible to make the grinding process easier. Allow the fill to dry and shrink, then go over it again.

Apply a second coat.

The finishing is done with various grades of diamond grinding pads, both power and handheld.

Continue with the tiles and other pieces in the same way.

Color-coding connotes different grits, from course to extremely fine.

Polishing

 The wet grinder is specially designed to work with stone. It has safety features to avoid electric shocks to the user.

Every now and then you may wish to rinse the piece to see your progress. The flow at the center of the grinder can be increased to a stream to wash away the excess materials.

 At the center of the pad there is a supply of water, controlled at the handle. The water flow should be a slight trickle, enough to keep the stone wet during grinding.

The edges should remain crisp, so keep the grinder flat and use special care on the edges.

Wet polish with a 400-grit pad. Be sure to keep the grinder moving. If it sits too long in one place, it will mar the finish.

A rinse with the hose reveals that some fogginess remains.

↓ Go over the whole surface by hand, again using care on the edges. After the hand grinding, the piece needs to dry thoroughly before a second coat of slurry is applied.

Note: In the first stage of applying the paste, the voids were rather large, requiring a heavy application. This tends to shrink and crack. This was followed by spot application. By the time we get to this second general application, the voids are much smaller and require a lighter touch.

↓ When the surface is completely dry, use the same "yogurt" mixture of slurry and apply a second coat to fill any remaining holes.

Let it dry. ↓

↗ Some of the smaller voids fill better by using fingers, so the basic technique at this stage is to rub it in by hand and scrape off the excess with a putty knife.

The next sanding uses a 600-grit diamond disc. ↗

 Use elbow grease on the apron.

 Do the same at the corners.

 Remember the underside.

 Finally go over the whole piece with an 800-grit grinding pad. Let the counter thoroughly dry before sealing.

 Ease the edges to prevent chipping. Start with a heavier grit such as this 200-grit, and work up to the finer grit.

Apply penetrating sealer with a spray bottle or mist application, and wipe away excess with a clean cotton rag. When the penetrating sealer has dried, go over the surface with an ultra-fine Scotch Brite™ pad. See page 68 for more information on sealing.

The most important thing to tell you about the hard-trowel technique is to be PATIENT. When pre-casting you can come and go, but when casting in place, fabricators onsite often err because they hurry the steps. Always start early in the day and remind yourself that patience is the key to success.

When casting a non-pressed finish, the steps I use are similar to pressing except you are casting right-side up instead of upside down. You will vibrate the edges after you pour the mix in the mold, which you don't do with pressed. But you'll still use reinforcement, keeping it toward the bottom of the slab. The methods differ in the finishing. Finishing a hard-trowel countertop is a process of making the counter flat and bringing up the cream as the bigger stones drop down in the slab; continuing to steel trowel it at intervals throughout the day with decreasing sized trowels; and flattening and burnishing the surface throughout the course of the day. You're putting a skin on it and it is drying slowly, nurtured with attention and care. You can see why I say start early in the day. Sometimes we cast a slab at nine in the morning and finish around two in the afternoon. That's about an hour between each trowel session.

Too much burnishing causes "chattering" from the trowel, so strike a balance between not enough and too much effort. If you don't leave it alone, it won't firm up. Don't stand there and trowel all day; leave it before it gets overworked. I use stainless steel rather than magnesium trowels because magnesium darkens my white cement.

Use a figure eight motion that makes a flowing trowel line, holding the trowels slightly on edge once the slab starts to harden. By the time you get to the smallest "midget" trowel, you can use more pressure.

Spray a little water on the trowel (not on the slab which can discolor it) for lubrication if need be. Be careful not to dig into the creamy surface.

Know when to stop, but don't stop too soon. It's like painting a picture in that way. Wait at least overnight, preferably two days, to hit the slab with #220 sand paper just to take off any burrs or nibs.

I believe that some trowel marks show the hand of the craftsman, and many clients like them. A light sanding shouldn't remove the trowel marks. I don't polish with diamond pads on hard-trowel surfaces because they will remove the top skin and the crafted appearance.

Casting

To see mold building for a hard trowel and ground (right-side-up) project, turn to page 70.

 Begin by applying a layer of concrete about one half the thickness of the slab to the melamine surface. Work it up on the edges of the mold walls.

After carefully positioning sink and faucet knock-outs as before, pressing the mesh into the concrete with your hands first. Wear rubber gloves.

 At the same time you can fill the molds for the wall tiles.

Use a scraper to help embed the mesh into the cement.

 Add the first level of metal mesh.

When the mesh is well seated and covered with concrete, lay in threaded rod in front of and behind the sink.

↓ Add another layer of concrete over the whole countertop, and then put the second mesh screen in place.

It is now time to make the first smoothing with a ↓ wood float.

↓ Work the mesh into the concrete and add more concrete to fill the mold.

Work the whole surface, filling in any voids as ↓ you go.

↗ In the narrow areas around the sink and fixtures, apply extra pressure to compress the concrete and remove air bubbles.

Use the screed or wood float on the wall tiles in ↗ the same way.

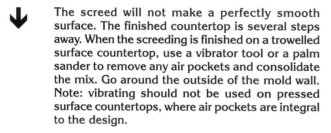

The screed will not make a perfectly smooth surface. The finished countertop is several steps away. When the screeding is finished on a trowelled surface countertop, use a vibrator tool or a palm sander to remove any air pockets and consolidate the mix. Go around the outside of the mold wall. Note: vibrating should not be used on pressed surface countertops, where air pockets are integral to the design.

Do the same thing on the backsplash tiles.

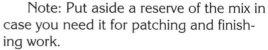

The countertop is left alone to set-up before the next step. We are waiting for the water or "cream" to rise to the surface. This will take an hour or two, depending on temperature and humidity conditions.

Note: Put aside a reserve of the mix in case you need it for patching and finishing work.

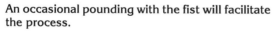

An occasional pounding with the fist will facilitate the process.

 There is a choice to be made between fiberglass and wood trowels. The fiberglass is preferred for this fine work because it doesn't warp. This assures a nice flat surface.

The basic trowel motion for leveling is a figure eight.

 A small amount of concrete is used to fill in voids as necessary.

Work your way around the countertop.

 The leveling or screeding process begins by working in from the edges. The mold walls act as guides to assure that the countertop is level.

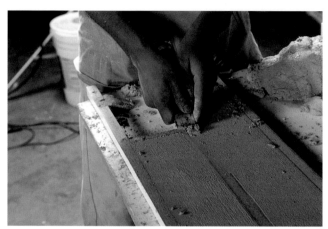

↓ Pay close attention at the corners. You want them to be nice and crisp.

Note: Before leveling, it is important to clean the top of the mold wall. For this job we have taken a corner out of a putty knife. It leaves a 3/4-inch tab from a 1-1/2-putty knife that acts as a guide on the walls, including the inside walls of the tile molds.

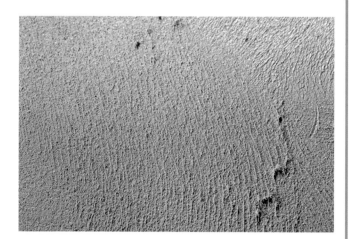

↗ The resulting level surface will look textured. Let it sit for some time. It is important not to rush the finish. Working the concrete brings water to the surface. This is the "cream" that will allow for a smooth, final trowelling with the metal trowel. By letting the concrete sit between steps you will always be working on an established surface.

After cleaning the wall edges, remove the waste with a trowel.

 When the surface is firm to the touch, use short strokes to bring up the cream, and long strokes to smooth it out. In this case I am using a good amount of pressure. This is possible because the concrete has set up considerably.

Finally, smooth it over with the trowel.

Occasionally there will be an air pocket in the surface. Cut into it with the trowel blade.

Here we need to remedy a slight dip.

Then knead out the air pocket with your fingers.

Add some concrete to the area from the reserve, then smooth with the trowel.

↗ We are getting to the point where the countertop is level. The strokes of the trowel are still visible at this stage. The cream is still rising, so we will let it sit for another fifteen minutes to half an hour.

After a final trowelling, allow the concrete to cure for at least 24 hours before proceeding.

Finishing Hard-Trowel Project

↓ After the countertop has cured for at least 24 hours, elevate the top on blocks and pop out the sink mold. A couple of light taps should knock it down and allow you to remove it.

The handle of a putty knife can be used to pop out the plumbing fixture forms.

↗

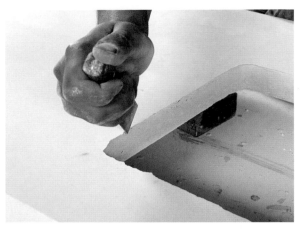

 A rubbing stone and putty knife are good tools to clean up the sink opening.

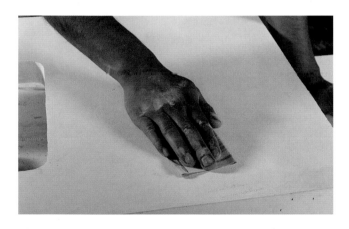

 By hand, dry-sand very lightly with 220-grit paper over the whole surface to remove burs. We are preserving the "hand of the trowel." Orbital sanders and other electrical sanders are far too aggressive for this job, and will quickly expose the aggregate. The edges are left crisp until after sealing. Then a paste will be used to fill any voids and to clean up any damage done to the edge.

A close-up shows the bug holes left after curing. These are remedied with colored cement paste.

 Brush the surface clean.

Before filling the bug holes we apply a preliminary coat of penetrating sealer.

The penetrating sealer can be applied in a number of ways, including foam or bristle brush. I think a hand-pump sprayer is the best application method.

After you get a color match, apply the paste by gloved hand, filling the larger holes first.

Using a clean rag, remove the excess sealer and even out the application.

Next go over the whole piece with a very thin coating, working it in and rubbing it off. As you rub and apply it, it will almost dry to your touch.

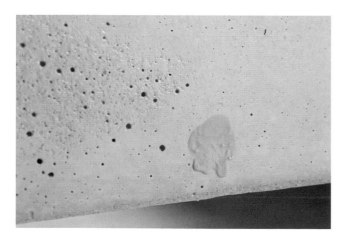

BRCP products include matching dry-powder pastes, or a paste can be created using a mix of Portland cement, concrete bonder, original color, and water blended to a yogurt consistency. Apply a dab to the countertop. If it is too dark, as shown here, add more Portland cement. If it were too light, I would have added more color. Both of these adjustments should be done in very small increments. For this 14-square-foot countertop, we should prepare about two cups of paste.

After allowing the paste to set overnight, sand by hand with 220-grit paper.

Using an 80-grit diamond pad, round or "ease" the edges of the sink and the finished edges.

Sealing

The look I have always gone for is a light satin sheen, not too shiny. I don't see the point in creating an earth-hewn, hand crafted look and then covering it with a plastic coating. The counters fabricated for the apartment photographed for this book were sealed with BRCP Penetrating Sealer and three applications of BRCP Satin Sealer, applied with a foam brush. An alternate method of applying sealer is simply with a clean soft cottton rag as shown.

We apply one coat of penetrating sealer by misting with a floral sprayer or wiping with a clean rag lightly saturated with sealer. Don't let it puddle; wipe off any excess, and let it absorb. When the slab is dry to the touch, in about an hour, the Satin Sealer may be applied.

The Satin Sealer is applied with a foam brush lightly saturated on the tip with sealer, or with a clean soft rag. Each application method is repeated once or twice, with about twenty minutes in between, or until the surface is dry to the touch. If rub marks or brush strokes appear, they can be buffed off with a ScotchBrite™ pad. Let the last coat dry overnight. We usually wax the slabs after they are installed, that way they are easier to carry and less likely to slip out of our hands on the way up the stairs!

A grinder pad adds flexibility for easing the edges in the corners.

Clean the surface thoroughly with clean cotton rags.

Apply satin sealer with a sponge brush, as shown, or a clean cotton rag. Use a light hand or you will leave brush strokes. Cover all exposed surfaces, including around the sink, the faucet area, and the front edge. Let the last coat dry overnight. We usually wait to wax the slabs until they are installed. That way they are easier to carry and less likely to slip out of our hands on the way up the stairs!

A Ground Surface Project

Like the trowelled surface countertop, the ground or polished top has a very smooth finish. The mold is loaded in the upright position, vibrated to insure a minimum of air pockets, and polished to a fine finish.

The ground countertop mold is formed right-side up. Polishing can be done lightly or aggressively to reveal the aggregates or larger pieces of stone or other decorative materials combined with the cement mix.

Like the hard-trowel counter, I cast the slab for a ground surface right-side up. Although it can be poured upside down and vibrated, I cast it right-side up to control air holes and color consistency. The first important step after filling the mold is getting a flat surface. I use a figure eight motion with a wooden trowel to "float" the surface before I start the finishing.

Decide how much aggregate you want to show. If you really want to show a lot of aggregate, take down the top 1/16-inch layer the day after your pour, with a #50 grit diamond pad on a water-fed polisher or a dry grinder, before you start the finishing polish. My mix uses a marble chip aggregate that I can choose to highlight with an aggressive grinding. Most often, I want

a lightly ground surface which requires a progression of diamond pads by grit number, after 3 days of drying.

Starting with #200, move through #400, #800, #1500, and finish with #3000. If holes pop out when you're grinding, fill them with a slurry paste.

The day following your cast, lift the slabs onto 2 x 4 blocks. After stripping the forms, fill any edge holes with a bit of cement paste or slurry before you start grinding. Always pay attention to finishing the underside, ends, and edges of the counter, as well as the top. These details distinguish your craftsmanship and make the difference between a mediocre "sidewalk" type counter and a skilled artisan's work.

A Note About Aggregate

There is a whole world of aggregate possibilities out there: seashells, glass, iron filings, aquarium rock, quartz, and obsidian. Experiment! I made a set of countertops for a restaurant in San Francisco with hammered shells the owners brought me from their oyster farm on Tomales Bay. I've added pieces of CDs, tiny plastic toy combs and brushes for a vanity top, colored rock, and marble chips. It helps if the aggregate has rough edges, but it's not essential. I don't use any large aggregate in my bag mix, but any aggregate should be assorted in size, and you can add fine to 3/8-inch aggregate and larger if you choose.

Building a Right-Side-Up Mold for Hard-Trowel or Ground Finishes

Mold making is one of my favorite parts of the process. Like a puzzle, I find it rewarding to think through. If you can cut wood straight with a table or skill saw, it is pretty simple. Although it's daunting when you make mistakes, the fun part is figuring out how to make the project correctly.

For simple straightforward kitchen slabs, I use melamine to build molds. It is relatively inexpensive and has a smooth hard surface cement doesn't stick to, and it is reusable. Because melamine has two sides, I can use the opposite side when the first gets tired looking.

Unlike Formica, melamine is flat rather than rolled, and it doesn't require laminating it to plywood, an intensive process that I stopped bothering with years ago. Some fabricators use Plexiglas or tempered glass. There are lots of options. Rough surfaces too can give a unique surface for specialty applications. I chiseled melamine once for molds to create a rocky look for the Illuminations stores.

Once all the decisions have been made, you get to go at it. Step by step, build melamine sides up against the Luan template, nailing the template down to the surface with a finish nail so it doesn't move around and you can push against it. (Luan is a cheap grade of plywood available in 1/8" thick sheets, usually 4' x 8'. We cut the Luan into 2 1/2" wide strips to use when templating.) You do need level and stable tables. Six legs are ideal because they support the middle. Remember, if a table has a twist in it, the slab will twist too. If the table sags, so goes the slab.

Remember to check and double check. If you're lucky enough to have someone else around helping, it's a boon to have a second set of eyes checking what you might not see in the midst of the process. Mistakes are only made because you can't see them!

Tack the template in place. In this case, because the finished surface will be on top, the melamine surface can be used or dirty or scratched.

Dry fit the mold walls around the template. To save cutting, the long pieces extend past the ends of the template.

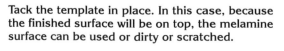

 Fit the crosspieces.

Holes are placed every 15 inches or so.

One of the reasons we use a template is that no countertop is exactly square. This end of the counter top is 1/4-inch wider than the other. With the template you can easily adjust for this.

Mark the centerline of the sink on the back wall of the mold to locate the sink knockout.

Drill through the walls using a countersink drill bit. The screw heads must be below the surface.

Back Splash Mold

 The backsplash will be made up of large tiles. They are 3/4-inches thick, so the mold wall will be 3/4 inches high. It can be made on the same large sheet of melamine as the vanity top. In the kitchen we built a template for the backsplash, because it contained electrical outlets. In the bathroom backsplash there are no outlets, so the mold can be built based on measurements from the countertop.

Using a straight edge to establish an edge, screw ⬇ down one wall of the tile mold.

Chop several pieces on the chop saw. The easiest ↗ and most accurate way to do this is to create a stop at 4 inches.

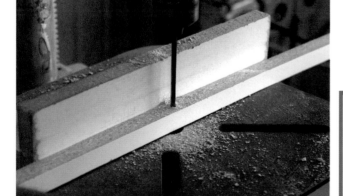

↗ Predrill holes in the edge of the tile mold – a drill press makes short work of this. Countersink the holes. The screw heads must be below the surface.

Aside: Sinks and Faucets

Sometimes the designer of the counter space doesn't take into account the space needed for the sink and faucets. It is important to make sure that the sink will fit into the cabinet as shown in the drawing. You might need to move it forward or backward. I like to bring the sink back to the shop if possible. New sinks in the box often have a paper template with measurements of the sink, which is very helpful to have when making the mold.

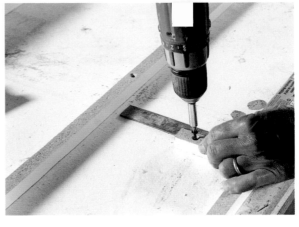

To determine the width of each tile, I divide the overall width of the vanity top by three. Since the vanity top is in an alcove, I will also create separate tiles for a return at each end, for a total of five tiles. For the return tiles, remember to subtract the depth of the backsplash tile, or 3/4 inches.

Using a square for positioning, screw the next mold piece in place. Continue in the same way for the third and fourth mold pieces.

Screw the end piece in place.

With the dividers in place, align the top edge using a straight edge and screw into place.

Measure the length of the tile, and then drop back 1/16 of an inch to allow for the seam between the tiles.

The same method is used for the sidewalls; use the bottom of one tile mold to form the top of another course.

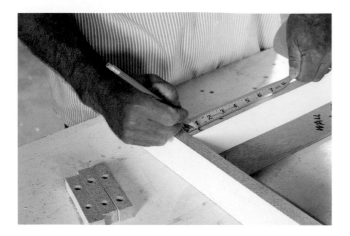

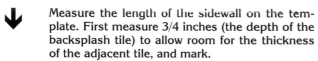 Measure the length of the sidewall on the template. First measure 3/4 inches (the depth of the backsplash tile) to allow room for the thickness of the adjacent tile, and mark.

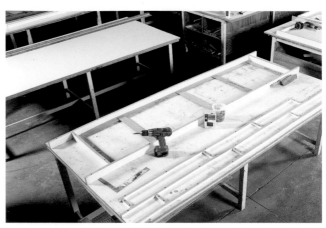

Now we have the basic molds for the vanity top and backsplash.

 Then measure from the other side of the template. This return will allow for the 3/4-inch thickness of the backsplash tile.

Although we want the countertop to appear to be 2-1/2 inches thick, in reality only the front edge will have that thickness. This is done by filling the mold with 1-inch-thick Styrofoam, leaving a 3/4-inch gap at the front edge for the apron. Measure the depth of the countertop.

HINT: It takes very little effort to create an extra piece, in case one breaks. Make it the length of the longest tile and add it to the molds for the returns.

Cut the Styrofoam to the width of the vanity on the table saw. This gives a clean cut. Take care – Styrofoam tends to pull quickly through the saw.

Remove the template from the mold.

Use a 3/4-inch melamine spacer on the front edge to create space for the return edge.

Align the template on the foam and scribe the end with a putty knife or sharp blade.

Tack the Styrofoam in place. There may be a slight gap at the back due to the imprecision of cutting Styrofoam. This will fill with concrete, but can easily be broken off.

A couple good scribes and it should break off cleanly.

Because one side was wider than the other, we add filler Styrofoam to the gap in the back.

Important Details

Making a Sink Knockout

I visited some people near my cabin in the mountains recently and found they had a Buddy Rhodes countertop, circa 1986 or so, which was purchased by the previous owner as a whole slab and a hole was cut onsite for the sink. It's been a long time since I did that, though it can be done. The cut-to-fit approach to countertops works for a drop-in sink, but an under-mount sink surround requires a finished edge.

A knockout is where you cast around a place left for a sink, and knock the foam out after the concrete has set. I use Styrofoam for knockouts. I tried wood, but found that the concrete shrinks and the wood knockout swells. Alternately, foam will give with the shrinking concrete.

If you are using a new sink, employ the paper template supplied by the manufacturer and found in the sink box. The paper template can be spray mounted onto Luan, from which you can make a rigid template. From this pattern you can make a foam knockout as pictured in this section. An actual sink can similarly be outlined onto Luan.

Remember to make the opening of the slab 3/8 of an inch smaller than the sink outline, as it looks better to extend over the edge than trying to line it up exactly at the perimeter.

Wooden dowels for faucet holes don't work for the same reason as wooden sink knockouts. Use foam or rubber or PVC pipe slit vertically and taped so it has spring to it and somewhere to move as the concrete shrinks.

Use a spray-on adhesive on the back of the sink template.

We need to fit the countertop for the undermount sink. First I cut out the template provided by the company.

Apply the template to a piece of luan.

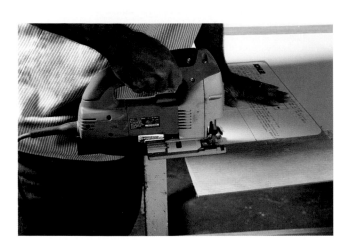

 Cut around the template with a jigsaw. I leave about 1/8-inch of wood showing.

With a jigsaw, trim the Styrofoam so it is close to the edge of the sink template. The jigsaw leaves a rough edge, so you want to leave some extra Styrofoam for sanding.

 Sand the luan down to the template.

At the sanding wheel, gradually remove the foam. The plywood backing not only gives it rigidity, it also allows you to hear when you have removed enough Styrofoam. The right angles created by a platform feed are a very important result.

Tack the template to a piece of 1-1/2-inch Styrofoam cut roughly to shape, but larger than the template.

The result.

 Mark the centers.

 Remove the template and luan.

 Draw lines through the centers.

Taping the edges of the Styrofoam gives a smooth, clean surface to the concrete. I use a thick, two-inch wide tape that has some body to smooth out the foam. This makes it easier to see bubbles and other imperfections, and the stiffness helps it hold up to the pressure of the concrete. Start at the centerline of the long edge and apply the tape all around.

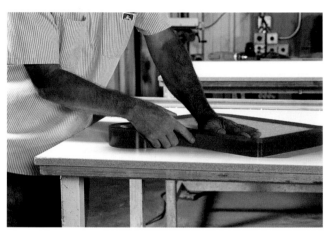

Continue around the piece until the tape meets.

Cut the tape flush so there is no overlap.

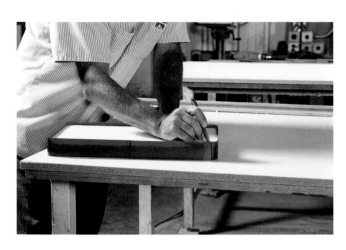

Before folding the tape over, make small cuts on the curved corners to enable you to get a smooth corner.

This edge tends to get caught up in the trowel, so I cover it with another piece of tape.

Fold the tape over, starting with a straight edge and working your way around.

Returning to the countertop mold, draw the centerline across the Styrofoam.

At the corner overlap the small tabs to get a flat turn.

With the tape seam toward the front edge of the vanity, align the centerlines and place the sink template 4 inches from the front line (or whatever your plan calls for).

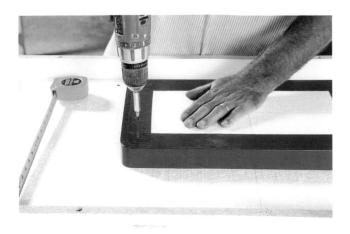

⬇ Screw it in place. The screws should be long enough to go into the melamine, but not through it.

⬇ Create a pilot hole for attaching the plumbing placeholder. Wrap the plumbing fixture placeholder with tape to facilitate release after the concrete has dried.

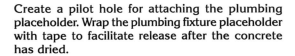

⬇ Use a straight edge to make sure the sink mold comes to the top of the mold wall, but not above.

Place the screw through the rubber knockout and find the pilot hole. The large ring of the plumbing placeholder should always be on the underside of the counter. On the bathroom fixtures, which are molded right side up, the large ring of the plumbing placeholder should be down against the foam. ⬇

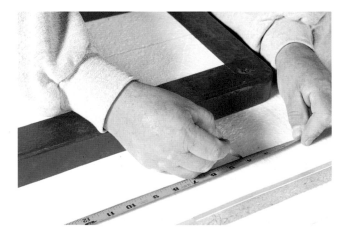

↗ Locate the centers for the plumbing fixture.

Now that the mold is complete, it is time to ready the reinforcement. In narrow, weaker spots like the spans in front of and behind the bathroom sink opening, galvanized or hot-dipped threaded rod is measured and will be inserted later for reinforcement.

Reinforcements

Reinforcement choice is very important. When I first started experimenting with countertops years ago I cracked many a project by using re-bar. Re-bar may work in a two-inch or thicker countertop if it is placed in the bottom of the slab, but I don't recommend it. Re-bar can cause cracking in thinner countertops because the cement naturally shrinks and the re-bar doesn't. Because re-bar is not galvanized, it rusts and consequently expands over time, causing the concrete to deteriorate.

Because I make countertops that are typically an inch and a half thick, I have had to choose reinforcement that is a thin gauge, yet strong. Although I built my reputation over the years using galvanized wire mesh as reinforcement in all of my countertops, I also recommend Dur-O-Wal™ ladder wire, which I have been using more lately (see sidebar). You can use four- or six-inch welded wire as well. The mesh is particularly good for molds that go vertical (planters and furniture), as it creates a sheet to hold up the cement. It is also well suited for the pressed finish.

My colleague, Jeff Girard, a civil engineer and founder of Concrete Countertop Institute, teaches his students that concrete countertops should be seen as structural beams rather than slabs on grade. Reinforcement is critical when dealing with structural beams. He recommends ladder wire.

You can use zinc-plated threaded rod, or 3/16-inch stainless steel rod around sink openings.

Whichever reinforcement you use, keep it in the bottom of the slab rather than the middle, a practice that helps prevent warping. Recently I have seen hardware cloth used, and the latest hi-tech approach is a carbon fiber material. For now, I'll stick with ladder wire or expanded wire mesh.

Ladder Wire

Ladder wire is used in masonry and glass brick fabrication. The brand pictured is called Ladur™ reinforcement and manufactured by Dur-O-Wal™. It is prefabricated structural reinforcement especially designed for embedment in horizontal mortar joints of masonry. It is easily adapted for placement between layers of concrete mix packed into a mold as shown.

Reinforcements/Fiber

I stopped using fiber in the early days of experimenting with concrete countertops because I didn't like the fibers showing and a hairy countertop just wouldn't do. But they can really strengthen a slab if used in the backing layer rather than near the surface. Mix in a handful of polypropylene fiber for every 70-pound bag of cement. It holds the cement together, providing tensile strength. Expanded wire mesh or ladder wire and fiber make an excellent reinforcement combination. I do not always use fiber each time I make concrete slabs, as it is expensive and not always necessary; but I always use it when packing sink and furniture molds with wire mesh reinforcement.

Installation

One of the most important stages of installation is transportation. Ease of transportation is related to the weight and length of the slabs. For this reason, the same people ideally do both templating and installation.

When my installation team goes to the site to template, they determine the best access to the installation site. If we can back the truck up to the cabinets we will most likely make the pieces longer than if we have to carry them up three flights of stairs! Should the client desire the hefty look, we can cast slabs an inch and a half thick and roll the edge, called a return, to make it look three inches thick. That makes it easier on both the installers and the beautiful new cabinets. We strap the countertop slabs onto a wooden "A" frame, which is placed on the back of a flat bed truck. We carry each slab like glass on its spine, the same as is done with marble and granite.

Other key installation decisions besides slab size are made during templating. It is important to plan for seams at the sink with pre-cast countertops. We usually make rails front and back and pin them in place with threaded rod and epoxy, rather than making a slab with a hole in it. Cracks are most likely to occur around the stress points of a sink opening and seams around the sink prevent that stress.

Dry fit the slabs in place to ensure that all has gone well in the mold-making process and to determine if any adjustments will need to be made on site. These might include notching, as in the kitchen shown, or grinding to ease the fitting. Once the dry fitting has gone well, the placing and gluing process proceeds.

For the grout between the slabs, we make up a colored caulk by adding the same colorant used in the cement mix to a white sanded caulk and stuff it into an empty caulking tube. We leave extra at the job site for touch up jobs or future re-caulking as need be. Finally, we always leave the client with a container of BRCP food-grade Beeswax and suggest that they use it regularly to maintain their counters.

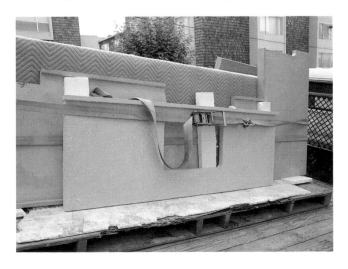

↗ Use an A-frame with belts, Styrofoam, and blankets, and carry the countertop on the edges, like glass or stone, to make sure it arrives safely.

It's really important to handle the slabs delicately. The slab is carefully moved using foam barriers. ↗

 Holding slabs vertically end to end is the easiest and safest way to carry them.

Dry fit and adjust the corner piece. You can cut to fit if necessary.

Dry fit the right side of the slab.

 Here countertops have to be carried up two narrow flights of stairs. This was an important consideration in creating the original template.

Dry fit the back rail.

 Using the grinder with a cutting wheel, cut the outside corner and notch it to fit. Marking pieces with a pencil will make them fit snugly, as seen here.

 Using the saw blade as a grinding disc, smooth the cut edge.

Hone the rail edge to fit between the slabs. Note: the plywood sub-top is not drilled for plumbing.

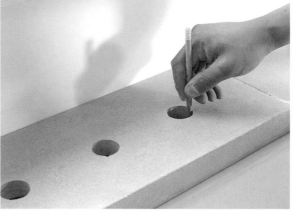

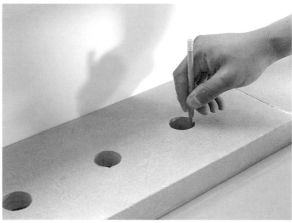

The hole spots are marked for drilling.

Using a circle saw bit on a drill, drill out the holes for plumbing fixtures as marked.

Take out Styrofoam slugs for electrical outlet box in wall.

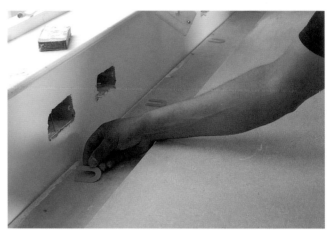

Place shims to level slabs.

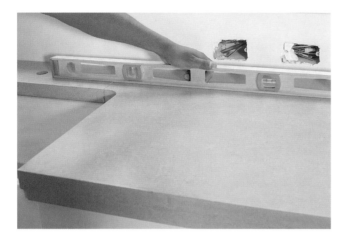

It is very important to level slabs on the cabinet before gluing.

Use shims to level the rail with the sink. A thin shim means a good fit.

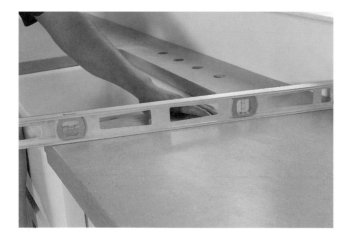

Level the slabs in the perpendicular direction also.

Place construction adhesive on the sub top with periodic puddles on corners and at regular intervals. The drying time is twelve hours. ↗

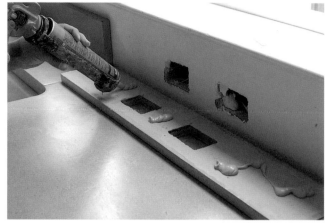

↓ Apply adhesive for rail slab.

Glue the backsplash using the same process. Place the backsplash firmly against the wall. ↓

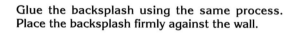

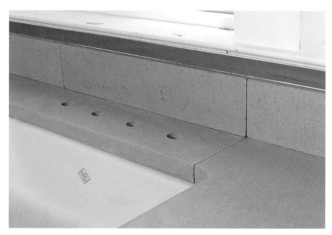

↗ Place the rail.

 Using blue tape will protect any exposed wall from caulk. Lift off as soon as possible to avoid pulling off paint. Tape can also be used on countertop at the seam to protect the countertop sealer. ↗

The caulk gun has been loaded with caulk of the same color as the countertop. Apply a bead of caulk on the seams and around the perimeter.

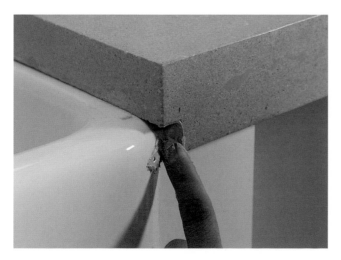

When caulking any notch, stray caulk should be cleaned off immediately using a wet sponge.

Maintenance

I have seen concrete counters that active cooks have used for ten years or more and they appear the same as when they were installed, and I have seen the opposite. Concrete counters can stain, but they are not significantly different from marble or other stones. Many plastic surfaces also stain; tile grouts discolor, and veneers can peel. The key to concrete counter longevity is care and maintenance. Concrete vitrifies over time; it becomes more watertight and dense as it cures and hardens. It is my impression that a ten-year-old slab of concrete is much less porous than a one-year-old slab. So careful maintenance is especially important in the first months of having a new concrete countertop.

The sealers and the beeswax combination I apply to countertops serves the function of giving the homeowner extra time to wipe up spills, but it is important not to let any food or liquid sit on the counter, especially oils and acidic food.

Regular waxing helps repel water as well as lubricating the surface. Cutting and heavy use areas should be covered with cutting boards, just as is the case with many other counter surfaces. Chips in edges can occur if heavy pots or objects bang them. They are easiest to repair in a pressed surface, but they all can be rectified. Knife slices and hot pans compromise the sealer rather than the concrete, but both should be avoided. Re-sealing, with or without sanding, can repair burns and worn or damaged sealer. Some of my customers routinely schedule re-sealing every few years. My own cast-in-place countertop, however, hasn't been sealed since I poured it in 1987. Although it has sustained many an assault of lemon fests from Margarita-making, for example, I just wax or oil it (olive works great) after the spills and the acid etched spots disappear. It is very dense from years of curing and oiling.

Troubleshooting

The most commonly experienced problems with concrete countertops are:

Warping
Cracking or uneven curing
Color variations
Errors in measurement
Incorrect templates

Most of the problems can be prevented by:

Curing the countertop evenly
Minimizing the water content of the mix

Reinforcing with the right material and placing it properly
Adjusting methods for cast-in-place applications
Working with good color samples (and a tight customer contract)
And thinking the whole job through before mixing the concrete

By thinking the whole job through before mixing, you can make important decisions that will prevent problems later. Ask key questions, such as, "Where will the sink go?" "What kind of back splash will work best?" "Have I made enough room for plumbing?" "Where will the seams go?" "How will I get the pre-cast slabs into the house?"

Don't Water it Down

When measuring the water for the mix, start with much less that the directions call for and add water until the consistency is right for the method being employed. I use four and a half quarts of water per bag of BRCP Mix for the pressed finish, and a half of a quart more for trowelled.

Casting in Place

Cast-in-place countertops are not my preferred method; however my friend Tom Ralston swears by them. When casting in place with my fine-aggregate mix, add 3/8-inch pea gravel and a handful of fibers per bag. This will lessen the shrinkage and make up for the uneven curing of counters spending their curing time trapped on a cabinet.

Aside: Molded Edge

I choose edge forms rather than working wet edges because I like the clean lines they produce, and I remove them only after the concrete dries.

Aside: One of a Kind Counters

Like a hand glazed pottery cup formed on a wheel and sold in a ceramics store, concrete countertops are hand crafted; therefore, color and texture may vary. Remember, your client should be seeking that earth-hewn quality. As far as color inconsistencies, I have largely avoided that since bottling my liquid colors to consistently produce a set palette of hues. If you experiment with the many possible color combinations, it is essential that you get a signed sample from your client. Your contract should state something like, "This is a handmade product. Color and texture may vary. Hairline cracking may occur." If your finished product matches the color the client signed off on, you are covered.

Cleaning Gray Water

Do not send the water you have used to mix cement down the drain. Any water used in the mixing process must be treated before it goes into the sewer or the ground.

What is needed in concrete water recycling is called a physical process, which means treating water that has pollutants suspended rather than dissolved. It is considered passive because you don't actually treat the water chemically or biologically. Rather, a system is set up whereby the water is allowed to sit for a period of time in a series of settling tanks, with water pumped from one to another after the silt has settled to the bottom of the tank. The sludge at the bottom is removed and dumped in an appropriate area.

I do not pretend to be an expert on water treatment; I recommend that you research this on your own, but the process is similar to that used by granite and marble suppliers and is documented widely online and in print.

I currently use five-gallon pails for grey water, and let the water sit and settle while the sediment sinks to the bottom of each pail. I siphon off the top water by pump and let it sit again. I do this three times before I let the grey water go into the sewer or into the ground. The sludge goes to the dump.

Having a plan for treating your water is a key component of the concrete fabrication process. Irresponsibility can put you out of business, or out of the neighborhood.

Gallery

This remodeled home in the wine country is filled with concrete counters and floors. The pressed floor tiles here vary in size and hue in an historic pattern. The counters and island color are trowel-finished with an iron oxide wash over an integral mushroom concrete base. The island close-up shows the hand craftsmanship prized by this homeowner, on her second Buddy Rhodes kitchen project.

This chocolaty "tone on tone" pressed finish holds up well with lots of large family activity. The bar is a pressed surface with an ash background color and earth infill.

Both pages: These lemony tone-on-tone pressed finish counters, designed by the homeowner, provide a raised eating counter adjacent to working surfaces.

Both pages: This homeowner is a great cook who spends a lot of time in her well-planned kitchen with a view of San Francisco Bay. The countertop was finished with a hand-trowelled method and stained a warm pumpkin tone.

Both pages: Striking concrete side panels give this central island its jet-age shape and appeal. While warm concrete texture and rich wood tones keep a sense of earthliness, a willingness to experiment with contemporary styling contributed to this unique and inspiring space.

Left: A concrete, cone table base matches the kitchen counter. This architectural element creates a hip, bistro atmosphere in a private home.

Below: A pressed tone on tone counter with an integral sink and drain board.

Blue laminate floor cabinets get an earthy grounding from a sleak, hand-trowelled countertop (in the ever-popular #14 color popularized by Pottery Barn). The sink is mounted under two rails, front and back, counter-balanced by a unique backsplash.

A generously sized farm house sink finds its place within a hard-trowelled countertop slab with no return edge.

A Seattle designer chose an eggplant tone-on-tone countertop for her own kitchen project.

The homeowner revels in this wonderful space, and bypasses the granite countertops enroute to gushing about "the hand of the artist" forever etched in the serving ledge side panels of the concrete island.

Both pages: Pressed table tops and hard-trowelled sideboard counters were installed at The Edible Schoolyard organic garden and kitchen program at King Middle School in Berkeley, founded by Alice Waters. Note that the students are working with cutting boards. Nine hundred students per year come through this kitchen and six years after installation the table tops are pristine, thanks to the care of project leader Esther Cook, who has the students work with cutting boards.

Warm earth-colored countertops work well in a traditional-style kitchen with wainscoted cabinetry. As lovely as this combo of lemons and concrete appear, try to separate acidic lemon juice from finished concrete surfaces whenever possible.

A strip of copper inlaid during casing creates a border in this pressed island countertop.

A hard trowelled taupe countertop
extends down the side of the outside
cabinet for a seamless profile.

The tone and texture of a hard trowelled countertop harmonize with the warm furnishings of a stylish kitchen.

A favorite with today's interior designers, an expanse of hard- trowelled, craftsman style countertop stained black adorns this handsome kitchen.

End panels, continuing the surface into a vertical element, are often used by designers to distinguish kitchens from the common approach to countertops. This peninsula counter doubles as a dining surface. The "Universe" finish in the BRCP color line was once called #14 and was developed for Pottery Barn stores.

A hard trowelled countertop was fashioned in a unique kidney shape for this distinctive island. It, and the curved countertops beyond, are in the Universe color.

A stretch of counter unifies a large kitchen composed of varying elements – antiqued green cabinetry, red tile back-splash, and multi-colored, playful touches from the sculpture to the lighting.

A pressed finish cone table and hard-trowelled countertops adorn a designer's kitchen.

This curved sink and counter was designed by renowned British Kitchen
Designer Johnny Grey for his San Francisco showroom.

The pleasing combination of stainless steel counters and a concrete
island can be easily viewed from above.

Concrete Concepts and Design, owned and operated by Buddy Rhodes associate, trainer, and former long-time foreman Heriberto Esquivel, fabricated this sky blue integral sink with a built in drainboard for Johnny Grey Kitchen Designs' San Francisco Decorator's Showcase House.

This signature Johnny Grey kitchen includes all his favored design elements, including custom colored concrete. Here pressed floor tiles and sidebar counters are coupled with custom, hand-crafted cabinetry and kitchen furniture.

This hand trowelled countertop was cast an inch and a half thick with no edge return. Its spare clean lines are mirrored in the adjacent glass tabletop.

A small galley kitchen embodies the earthy sentiments of its owner, from wood furnishings and a paneled ceiling, to the concrete countertops that create two long workstations.

Both pages: This apartment in Chicago has two Buddy Rhodes bathrooms fabricated in different years. The room shown has counters and tub trowelled with a cement slurry similar to the paste and acrylic additive we use to fill the veined surfaces. The vanity with integral sink was cast in one piece with an apron edge. The Japanese-style soaking tub was first built of a plywood sub-structure, plumbed, and covered with a water-proof membrane. We enclosed the sub-structure with panels of pre-cast concrete shipped to Chicago and installed on site. One-piece concrete bathtubs are tricky to fabricate, although they are done. Concreteworks in Oakland specializes in them. We find the paneling approach safest.

The counter with back splashes and a tub surround are hard-trowelled slabs in a caramel color. A built-in bench for the shower (not shown) matches the surround and counters.

The custom bath area includes a chartreuse countertop on a free standing cherry wood cabinet. The counter sills behind the bath are matching trowelled concrete.

White tone-on-tone tiles sheath this tub/shower
amidst the trees on a Marin county hillside.

Matching shower wall tiles are reflected in the mirror above
this vanity with a return edge.

Both pages: These two vanities in
a San Francisco interior designer's
loft show a black hard-trowelled and
a two-tone veined finish.

These aquarium-loving homeowners wanted an underwater theme throughout their project, reflected in their trough sink with wall panel back splashes.
The water flows down the length of the cast tiered trough.

Example of Hotel Healdsburg vanity counters and tub surrounds in guest rooms in the Sonoma, California wine country town.

These counters and the tub surround are washed with an iron sulphate stain in addition to their integral neutral color.

A concrete sink with a deep apron houses double overmount vanity sinks.

Moss and sand veined slabs were used as wall tiles in this bath with a stand alone glass vanity sink.

A Chicago restaurant vanity top is fabricated in pressed finish with an integral oval sink. Integral sinks are most practical in the upside down pre-cast pressed finish because it molds smoothly against the form without trowelling.

This bath in a Kohler, Wisconsin, showroom is
by New York-based designer Clodagh with stainless steel complemented by Buddy Rhodes universe concrete vanity and tub surround tops.

A gray tub surround provides a ground line to a beachfront wall mural. Hard-trowel surfaced panels clad the tub deck and platform.

Opposite page: Green was used for a custom countertop in a young girl's vanity, within a house where every bathroom boasts a different color concrete counter.

A vanity counter in a guest bath boasts a conservative moss green pressed finish.

Glossary of Concrete Countertop Terms Used by Buddy Rhodes

Resources: American Concrete Institute; Jeff Girard of The Concrete Countertop Institute, as published on Concretenetwork.com; and Buddy Rhodes.

A-Frame – An A-shaped wooden or metal framework used to transport countertop slabs on edge in a truck.

Aggregate – Granular material such as sand, manufactured sand, gravel, crushed stone, and blast furnace slag which when bound together by cement paste forms concrete.

Bug-holes – Small voids in concrete caused by entrapped air bubbles.

BRCP – Buddy Rhodes Concrete Products, LLC

Cantilever – A beam that projects beyond its supports or an area where a countertop overhangs a cabinet or support by more than a few inches.

Casting Table – A strong, level table designed for casting concrete slabs on top.

Caulk – Used to fill seams between countertop slabs. Usually color-matched. Made of a flexible material so as to create control joints in the concrete.

Cement (Portland) – The product obtained by pulverizing clinker (a hard mass of ash and partially fused coal that remains after coal is burned in a fire or furnace) consisting of hydraulic calcium silicates with calcium sulfates as an interground addition; when mixed with water it forms the binder in portland cement concrete.

Compressive strength – The ability of concrete to resist compression forces, or pushing together forces, expressed in pounds per square inch (psi).

Concrete – A material consisting of a binder within which aggregate particles are imbedded; in portland cement concrete, the binder is a mixture of portland cement and water.

Concrete countertops – A handcrafted alternative to manufactured countertop surfaces. Can be precast in a shop in molds built to the customer's specifications or cast in place by setting a form on top of the base kitchen cabinets and then filling with concrete.

Countersink – Drill recess for screw to enter, to keep screw head beneath mold surface away from sight and trowel.

Drop in sink – A sink that has a rim that fits over the countertop, also known as an over-mount sink.

Fibers – Tiny filaments made of polypropylene, polyolefin, nylon, polyethylene, polyester, or acrylic used to control shrinkage cracking. Fibers do not provide structural reinforcement.

Galvanized (Diamond) wire mesh – A woven mesh of wire strands, used as reinforcement in concrete slabs to control shrinkage and cracking.

Grinding – Mechanical surface preparation using rotating abrasive discs to remove cement paste or slight flaws and protrusions.

Install – Set a countertop onto cabinets, glue it down, and caulk any seams so that the countertop fits well and is level. Adjacent slabs are made flush and all fixtures such as sinks and faucets are properly mounted.

Integral sink – A sink made out of the same material as the countertop and forming a continuous surface with the countertop.

Knockout – A rubber or foam shape placed in a form where there will be a hole in the countertop.

La-Dur Wire – A galvanized steel wire product manufactured by Dur-O-Wal for masonry reinforcement, which can be imbedded in concrete countertops to provide structural support.

Luan – A cheap grade of plywood available in 1/8" thick sheets, usually 4'x8'.

Melamine – Particleboard coated with a plastic material. Often used in forming concrete countertops because of its smoothness and easy release.

Mortar Mixer – A mechanical mixer designed for blending cement based mortar. Mortar mixers use rotating paddles attached to a horizontal axle to mix mortar or concrete. Often used for mixing the small batches of concrete required for concrete countertops.

Penetrating sealer – A sealer with the ability to penetrate into the concrete surface to increase water repellency and resist stains. It does not change the surface appearance of decorative concrete.

Rebar – Reinforcing bars which can be installed in concrete slabs as a primary reinforcement to provide flexural strength. Rebar comes in various diameters and strength grades.

Return Edge – A countertop edge that is built into the mold to form a downward lip perpendicular to the surface of the countertop, making the countertop appear thicker without the weight or height of a thicker countertop.

Sealer – Liquid based material used to protect and enhance the appearance of decorative concrete.

Seam – A joint between 2 adjacent slabs of countertop material. Seams function as control joints in brittle materials such as granite or concrete.

Slurry – A paste used to fill bug-holes or voids in concrete countertops.

Shim – Pieces of plastic or wood used during installation to make slabs level and flush with adjacent surfaces.

Shrinkage – The tendency of the cement paste in concrete to shrink as it cures, causing concrete slabs to either curl (due to unrestrained shrinkage) or crack (due to restrained shrinkage).

Slab – A flat span of concrete.

Tensile Strength – the ability of concrete to resist tension forces, or pulling apart forces, expressed in pounds per square inch (psi).

Template – a physical pattern that represents the space into which a countertop will fit. For precast concrete countertops, templates are created on the finished cabinets, and then the templates are used to determine the size of the forms.

Under-mount sink – A sink that is mounted underneath the countertop.

Vessel sink – A sink that sits on top of the countertop.

Wax – Applied to concrete countertops for maintenance, enhancing the surface and providing short-term protection against stains but requiring re-application often.

The Buddy Rhodes All in One System

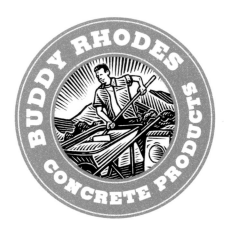

Most of the products used in the demonstration projects throughout this book are manufactured and distributed by Buddy Rhodes Concrete Products. They are available in construction supply and decorative finish stores throughout the country. A list of our current distributors may be found on our website: www.buddyrhodes.com.

Buddy Rhodes' pre-cast concrete its warm quality and versatility for over twenty years. The mix can be used bone white out of the bag, or with liquid color added to the mix water. Due to its rich consistency and plasticity, the mix can be used on vertical as well as horizontal surfaces, making it particularly suitable for Buddy's pressed technique and for working up mold walls.

One 70 lb. bag yields approximately 2/3 cubic foot or 5 square feet of counter at 1.5 inches thick.

Liquid Concrete Color

BR Liquid Color is diluted with a measured amount of water before mixing. One jar is formatted for two bags of BR Concrete Counter Mix. Because the liquid pigment is in the mix water, the concrete is colored uniformly.

Buddy Rhodes Concrete Mix

Buddy Rhodes Concrete Counter Mix (patent pending) is a refined combination of seven ingredients, including gradations of chosen aggregates, cement, and our own special ingredients that have given

Color Paste

BR Color Paste is formulated to match all BR Liquid Colors. The paste can be used as a slurry mix to fill air holes on any concrete surface. It was specifically designed to fill voids in the BR signature pressed finish. Voids may be filled with matching one or more contrasting colors for a desired effect.

Food Grade Beeswax

BR Food Grade Beeswax helps prevent stains and further enhances the color of concrete surfaces. It may be applied by the homeowner for regular maintenance over the life of the concrete countertop. Apply very lightly and buff it out with a soft cloth.

Natural Look Penetrating Sealer

BR Penetrating Sealer "natural look" is designed to impart water repellency, reducing water absorption. It is recommended for use in tandem with the BR Satin Sealer.

Acrylic Additive

BR Acrylic Additive is a liquid acrylic polymer emulsion specifically designed as an admixture for concrete projects. It is primarily used in lieu of water to mix the Color Paste. As an optional additive for the mix, it may be used to reduce shrinkage, promote curing, and improve abrasion and stain resistance.

Satin Sealer

BR Satin Sealer is a water-based and ultra violet resistant acrylic micro-emulsion. It will help protect the finished surface from oil and water-based stains and will enhance colors.

Sink Molds, Edge Forms, and Furniture Molds...

Buddy Rhodes Fiberglass Sink Molds

BR Sink Molds may be used many times in fabricating pre-cast integral sinks. Two vanity shapes and two kitchen shapes provide a range of options: oval, trapezoid, small and large rectangles.

RECEDING 1.5"TH   **PENCIL** 1.5"TH   **EXTENDED** 2"TH   **CHAMPHER** 1.5"TH

Buddy Rhodes Reusable Edge Forms

BR Edge Forms are made from extruded styrene and may be clamped onto a cabinet for molding the edging of cast-in-place countertops. They may be mitered on a chop saw for corners. Edge forms are sold in 8' lengths.

Buddy Rhodes Furniture Molds

Buddy Rhodes custom designed furniture molds take advantage of the unique qualities of the BR Concrete Countertop Mix: its pliability and capacity to be worked up a vertical surface.

Buddy Rhodes – Perles ME 140 Cement Mixer

Buddy Rhodes Concrete Products has teamed with Perles of Switzerland to supply a cement mixer ideal for mixing dry concrete mix in smaller quantities for concrete countertops. The mixer base harnesses 10.4 amps horsepower with two gears and a variable speed trigger. We provide two different paddles to cover a spectrum of mixing applications for cements, grouts, and mortars. Also included is a universal chuck for connecting after-market attachments. The package comes complete in its own rugged carrying case.

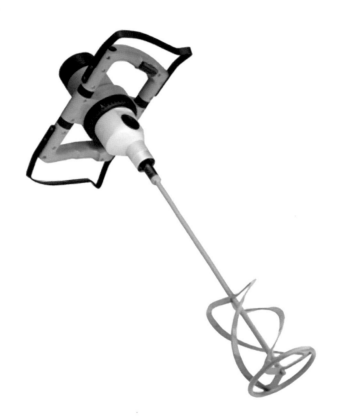

Buddy Rhodes Diamond Polishing Kit

After years of testing, Buddy Rhodes Concrete Products has introduced the ultimate polishing system for concrete countertops, taming the hardness of concrete and producing a very smooth luster. The Polishing Kit contains eight 5" diamond impregnated resin pads of graduated grits. The pads are designed to be used in sequence with a QRS backed flexible head attached to a water-fed polisher. We include 50, 100, 200, 400, 800, 1500, 3000, and Buff grits.

Photo and Design Credits

Photography

Doug Congdon-Martin: 98, 99, top of 132
Ken Gutmaker: 104, 105, 106, 118, 122, bottom of 123, 128, 130, 131
David Duncan Livingston: 96, 108, 109, 110, 111, 112, top of 123, 126, 127, 129, 135
Andrew McKinney: 97, 140, 141
Matthew Millman: 114, 119
Sohan Mutucumarana, 121
Sharon Risedorph: 100, 101, 102, 103
Caesar Rubio: bottom of 132
Wheeler Kearns Architects: 124, 125
Christopher Hirsheimer, leaf.

Designers

David Baker: bottom of 132
Sharon Campbell: 108

Concrete Concepts and Design: 121
DDL, 135
Glenn Dugas: 104, 105, 130, 131
Marian Elliott: 96, 126, 127
Gary Garman: 109
Johnny Grey, 119, 122
Katherine Lambert and Mark Kessler: 100, 101
Kelly Lasser: 116
Philip Mathews, top of 123, 129
Dan Phipps Architect: 134
Rupel, Geisler, McLeod Architects: 102, 103
Barbara Scavullo Design: 97, 114, 140, 141
Ruth Sorenko: 102, 103
Wheeler Kearns and Leslie Jones Interiors, 124, 125
Ana Williamson, 112
Wow Haus: 110, 111

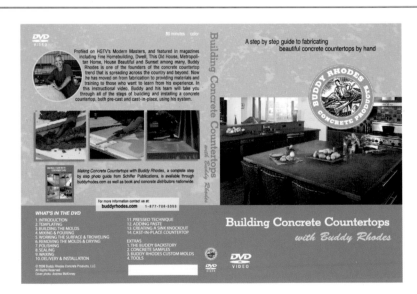

To complement the information in *Making Concrete Countertops with Buddy Rhodes,* BRCP offers an instructional DVD, Training Workshops at the San Francisco home base, as well as at distributors and satellite training centers nationwide. You may obtain telephone and internet support by contacting us toll free at 877-706-5303 or e-mailing us at *info@buddyrhodes.com*. The website is updated frequently with distributor locations, training dates and Tips from Buddy. We aim to support you in your concrete craftsmanship.